WHY DIDN'T I THINK OF THAT?

WHY DIDN'T I THINK OF THAT?

From alarm clocks to zippers

BY WEBB GARRISON
PICTURES BY RAY ABEL

Random House/ New York

To my grandchildren

Cover Illustrations/Charles Goslin

Published in the United States by Random House, Inc., New York.

Library of Congress Catalog Card Number: 79-87880
ISBN: 0-394-62021-6
This edition published by arrangement with Prentice-Hall, Inc.

Manufactured in the United States of America
987654321
First Random House Student Book Program Edition: September, 1979

Contents

Addressograph

During the 1890's some business firms employed hundreds of people to do nothing but address envelopes for the company mailings.

It was a long, slow and expensive process for the employer.

Joseph S. Duncan of Chicago decided to automate this addressing procedure. He typed each address in special rubber type and glued it to a piece of wood. Each wood block was stored in a machine, which could be inked, supplied with envelopes, and then would print the addresses one after another.

Duncan's "addressograph" worked, and he took out a patent on it in 1896.

Today, computerized systems are quickly replacing Duncan's invention. But he was the one who first replaced man with a machine for addressing envelopes.

Alarm Clock

Levi Hutchins of Concord, New Hampshire, was a clockmaker.

He was an industrious fellow who woke up at four o'clock every day so as to be sure to get a full day's work in.

As he grew older, however, he started to find it hard to get up when he wanted to. He kept oversleeping. It began to bother him a lot that he was becoming so lazy.

So he set about to invent a way to wake himself up.

He knew, from reading books, that the sun had been used to wake people, either directly or by fixing a mirror so that as soon as it rose, the sun's rays would reflect into the eyes of the sleeper.

But at four in the morning in New Hampshire, there wasn't any sun. So, that method did not help Levi.

Eventually he tried fitting a bell inside one of his clocks. It wasn't difficult to fix it so that the bell would ring at a specific hour.

So, Levi made the first alarm clock. He did not take out a patent on his invention, and thus received no money for the idea. But his clock woke him faithfully for sixty-eight years.

Ball Point Pen

Ladislo Biro lived in Hungary and worked as a proofreader, reading manuscripts when they were first printed and correcting any mistakes before the final printing.

He had an old-fashioned pen that he had to dip in a jar of ink before making a correction.

Ladislo didn't like that process. It was a bother, and there was always the danger of getting ink splotches on the page.

So he decided he would invent a pen that would not have to be dipped in ink.

He called it an inspiration. His friends called the idea plain foolishness.

Ladislo didn't worry about them. With his brother he tried many ways to make the new pen. Finally they hit upon putting a tiny metal ball at the tip, then filling the pen with paste-like ink.

That is the basic idea of today's ball-points. But it took many years to perfect the process.

When first manufactured, the new pens were very expensive, and few saw a good reason for buying them. Interestingly, it was the fighter pilots of World War II who popularized ball-point pens. Ordinary fountain pens leaked at high altitudes, and so the Air Force bought for the airmen the "high altitude writing sticks" that Biro made.

After that, ball-points became increasingly popular, and today many more are sold than fountain pens.

Band-Aid

Josephine Dickson married a man who worked for a company that manufactured gauze and adhesive tape called Johnson & Johnson.

We will never know the reason, but it is a fact that Josephine Dickson was accident prone. During the first week that she was married to Mr. Dickson, she cut herself twice with the kitchen knife.

After that, it just went from bad to worse. It seemed that Josephine was always cutting herself.

One day her husband had an idea. He sat down with some tape and gauze and a pair of scissors. Then he cut the tape into strips. In the middle of each strip he stuck a little square of gauze.

From then on, whenever Josephine had an accident, ready-made bandages were on hand for her to use quickly and without a lot of fuss.

At Johnson & Johnson, they heard about these new bandages that could be put on in thirty seconds. Soon the company was making them to sell on a small scale.

Four years later, in 1924, the company installed machines for mass producing the new product, and the trade name Band-Aid was adopted.

Bicycle

Back in the 1830's, grown-ups as well as children rode hobby horses for fun. There are not many left to see these days. A hobby horse is a wooden horse with a small wheel at each leg. The rider would push against the ground with his or her feet, and the hobby horse would roll along.

Scottish blacksmith Kirkpatrick Macmillan had an idea. It seemed to him that two wheels would be faster than four. Experimenting with the concept, he invented a system of propulsion based on pedals fitted to an axle on the rear wheel.

The new machine, based on the hobby horse, was a two-wheeled vehicle which the rider could pedal instead of push. It was called "bicycle," meaning "two wheels."

In 1840 Kirkpatrick pedalled his bike seventy miles in ten hours, an astounding achievement. Unfortunately, his ride ended when he knocked down a young lady by mistake and was put in jail for reckless riding.

The bicycle was a hit, however. Kirkpatrick was just twenty-four years old at the time. He lived to see his invention become popular in many countries of the world.

BLAKE, JONES &

Bikini

Not many of us would imagine there is a connection between atom bomb testing and bathing suits. But there is.

This is the story: in 1946 a French designer, Jacques Heim, made and sold in his beachwear shop what he called "the world's smallest bathing suit."

At the same moment, the American government exploded a series of atomic bombs from a site on a tiny Pacific Island called Bikini.

Somehow, advertisers were inspired to link the two events—perhaps because the amount of cloth in the new swimsuit was as small as Bikini Island.

At any rate, soon skywriters were advertising Heim's design, urging people to buy the "BIKINI—SMALLER THAN THE SMALLEST BATHING SUIT IN THE WORLD."

The new design, and the advertising, worked. Soon bikinis were seen at every beach and pool and became a standard item in summer wear.

Book Matches

Joshua Pusey was a well-known lawyer in Pennsylvania before the turn of the century.

He smoked cigars.

One day he was invited to a dinner party by the Mayor of Philadelphia. He dressed in his best clothes, and all was fine—except for one thing. The big box of wooden kitchen matches he was carrying to light his cigars, stuck out of his vest so much that he felt embarrassed.

Why did matches have to be so bulky? he wondered! Why couldn't they be made out of paper instead of wood? Paper matches would be lighter and much smaller.

Mr. Pusey worked at his idea, and in 1889 he patented paper matches. He tried to interest people in his invention, but for eight years, no one seemed to care.

Then in 1897, Pusey got his break. The Mendelsohn Opera Company wanted a special way to advertise their New York opening. They used books of paper matches with their name printed on them.

Suddenly everyone was talking about book matches, and paper matches began selling as fast as they were made.

In later years Joshua Pusey sold his rights to the Diamond Match Company. He was known to joke to his friends that he was by then known all over as "the match man," not as a distinguished lawyer who had pled cases before the United States Supreme Court.

Cash Register

Jake Ritty was a restaurant owner from Ohio taking a trip to Europe.

The year was 1879, before the age of air travel. Mr. Ritty was going by boat.

During the voyage across the Atlantic, the passengers were taken on a tour of the ship. What interested Jake most was something he saw in the engine room. It was a little machine that counted the number of times that the ship propeller turned.

"A machine that counts!" Ritty said to himself, and immediately thought of how it could help him in his business.

In fact, he was so excited by his idea that as soon as the ship reached Europe he came right back home to get to work.

In a short time, with his brother's help, he made a machine that added items and recorded the amounts. It was the first cash register, and he put it to work in his restaurant.

It made adding up bills and keeping records a great deal quicker and easier.

Ritty's first cash register was hand operated. Today's registers operate by electricity, but the basic principle remains the same.

95 00 5 10 15 20 25 30 35 40 45 50 55
$2 $3 $4 $5

Christmas Lights

In 1917, when Albert Sadacca was 15, there was a terrible fire in New York City.

It had been caused by a Christmas tree catching fire, for in those days Christmas trees were still lit by placing candles on the branches.

Now it just so happened that Albert's family, who had come from Spain, had a novelty business selling wicker cages with imitation birds in them that lit up.

Albert suggested to his parents that they begin making electric lights for Christmas trees. They had lots of bulbs on hand, and it would be much safer than using candles.

The Sadaccas thought Albert had a good idea, but only one hundred strings of electric Christmas tree lights sold in the first year.

After Albert thought of painting the bulbs red, green, and other colors instead of using plain glass, business picked up sharply.

Albert became the head of a multi-million dollar company.

COUNT YOUR CHANGE
HOME
OURNAL
XMAS TREE
FIRE KILLS 4
DELINEATOR

Chewing Gum

John B. Curtis was twenty-one when he got a job clearing brush midway through the last century. Like many foresters, he took to chewing the sap, or gum, that seeps from spruce trees.

But the gum lost its strength pretty fast. John tried cooking some on his mother's stove to make it last longer. Next he tried adding flavors.

John decided to sell his gum. He quit brush cutting and devoted his energies to "State of Maine Pure Spruce Gum," which he sold in small sticks, individually wrapped in waxed paper.

As a brush cutter John earned five dollars a week. In his first year in the gum business, he made five thousand dollars.

Crossword Puzzle

What is today's crossword puzzle came about because a newspaper editor had a problem. On December 21, 1913, Arthur Wynne, Sunday magazine editor for the New York *World,* was making up his paper. All the articles and advertisements were in, but on one page there was an empty space. How could he fill it? Wynne began to play with words on a piece of paper, fitting some into horizontal lines, others into vertical lines. Then he removed some of the words, which left empty holes. Here was his idea: he would ask his readers to guess what the missing words were. So that they could do this and have fun in the process, Wynne gave them a list of clues. Because words crossed one another, "across" and "down," he named his brainstorm the "wordcross game.

Now called "crossword puzzle," the game born of a newspaper editor's problem remains the favorite of millions. It is to be found in a corner of nearly every newspaper in the country, and scores of crossword puzzle books are published as well, so great is the demand.

Dixie Cups

In 1900, in the United States there were no drinking fountains. Instead, public drinking dippers in water barrels were widely used by thirsty travelers.

One young man, named Hugh Moore, decided the practice was unsanitary. He wanted first of all to make pure spring water available, and he wanted a way for people to drink it which would be germ free.

Moore designed a cup made of paper. These paper cups seemed clumsy at first, but a drink of spring water from a clean paper cup was so inviting that thousands tried it. Moore installed cup-vending machines at trolley line corners, and on a hot day a five-gallon bottle could be emptied in an hour.

Sales of the "Health Cup" were assured once Moore succeeded in having them placed in railroad cars. With the name changed to Dixie Cup, his sanitary paper containers became standard items on trains and in soda fountains.

By 1960, Dixie Cups were selling at the rate of more than fifty million dollar's worth per year.

EXCURS

Drive-In Movies

The 1920's and '30's were the golden age of movie-going in the United States, and movie houses were usually packed full.

One person who had worked in the movie theater industry, Richard Hollingshead, got an idea for an alternative to the crowds and stuffy theaters. Why not build a drive-in theater where people could watch the film on a giant screen from the privacy and comfort of their automobiles? he thought.

RCA-Victor agreed to build Hollingshead's drive-in. It was a big engineering job, and from start to finish construction took three years. Located on a ten-acre plot in New Jersey, the first drive-in motion picture theater opened for business on June 6, 1933. From their cars the first customers saw the movie from a screen fifty feet wide and forty feet high. The parking lot sloped gradually downward toward the screen for good viewing.

At first drive-in movies only operated during the warm months, but the development of individual speakers and in-car heaters made them a year-round institution.

Electric Fan

The electric fan was developed by Schuyler S. Wheeler, a highly educated scientist who is known for over one thousand inventions.

Many of his inventions consisted of ways to use electricity for the service of man. The fan is a good example.

Before Wheeler's time, people used hand-held fans to keep cool during hot summer months. Wheeler got the idea for a motor-driven fan while waiting for the Staten Island ferry in New York City in 1882. He noticed the way the ferry's propeller made expanding waves in the water, and thought he could use the same principle to create expanding waves of cool air.

Wheeler perfected what he called the "buzz fan" in 1886, but it was not until electric motors were mass produced some years later, that electric fans started to be widely used.

Now Wheeler's invention has become an old-fashioned way of staying cool because most people use air-conditioning.

Elevator

Inventing new products and machines is sometimes a risky business. Chemicals explode and vehicles crash.

In the case of the elevator, Elisha Otis is famous not for the original idea, but for figuring out how to make it safe.

Throughout history men have used their ingenuity for lifting heavy objects. In the 1830's, elevators were developed in both England and France for lifting persons as well as freight. They were not popular, however, because they were extremely dangerous. It was not infrequent for an elevator cable, under a thousand pounds or so of pressure, to snap.

What Otis did was invent a safety device utilizing a wagon spring which held the elevator in place if the cable broke.

Otis demonstrated his device at the Crystal Palace Exposition in New York in 1853. In front of a large crowd, he rode up in an elevator. Then he cut the cable. His emergency device prevented disaster, and he remained aloft, waving his top hat at the people below.

Elisha Otis' safety elevators made the modern city possible, with its skyscrapers rising a hundred stories in the air.

Eraser

People have been writing with pencils for centuries. Erasers are not new, either. Pieces of what was called "India Rubber" were used to remove unwanted pencil marks.

India Rubber erasers worked just fine, but they kept getting lost under a pile of books or falling off the table. You could never find an eraser piece when you needed it.

That was the situation until H. L. Lipman of Philadelphia, Pennsylvania, tackled the problem in 1858.

It was simple. Lipman took out a patent for a pencil with a groove at one end, into which a bit of rubber was cemented.

The eraser never had to get lost again.

And Lipman was never poor again. His patent earned him $100,000.

Ferris Wheel

In 1892, to celebrate the 400th anniversary of Columbus' discovery of America, a huge world's fair was planned. It was to be held in Chicago, Illinois, to be called the World Columbian Exposition.

Engineers all over the country were asked by Chicagoans in charge of buildings to offer designs for an attraction that would compete in originality and size with France's Eiffel Tower.

A 31-year-old engineer named George Washington Gale Ferris set to work, dreaming of an enormous revolving wheel where people could sit in cars attached at the ends of the spokes, and be whirled in a circle through the air. People laughed. But Ferris constructed his steel-spoked wheel, with thirty-six cars, each carrying forty people, and it operated on the midway at the fair. Riders were scared and thrilled at the same time, and all arrived back safely on the ground. The wheel drew the crowds. The original wheel was eventually sold for scrap, but smaller copies of it turn breathtakingly at fairs and amusement parks all over America. Young Ferris' wheel is gone, but his name lives on.

Flashlight

The idea of a lighted flower pot may strike us as pretty humorous, but it was the answer to a lifelong struggle for a Russian immigrant named Conrad Hubert.

Hubert came to the United States in 1890 when he was about 35 years old. He was flat broke.

He did what he could to earn a living. He worked in a cigar store, ran a restaurant for a while and managed a boarding house. He even tried farming and repairing watches.

Whatever he did, however, he never made much money. All he wanted was to stop worrying about making ends meet.

Now Hubert had a friend named Joshua Lionel Cowen who was very interested in electricity. Joshua had invented a flower pot with a battery in it. Electricity from the battery made the flower "light up" when a button was pressed.

Hubert decided he would try to sell these flower pots. Meanwhile, Joshua became interested in something new—electric trains—and he sold his friend the flower pot idea for almost nothing.

Fortunately, Hubert never went ahead with the flower pots because he had had an idea for a modification. He took the battery, the bulb and the paper tube from the pot and remade it into what he called "an electric hand torch."

Hubert sold his invention at first as a novelty, but the usefulness of the flashlight soon became apparent.

When he died in 1928 it must have seemed to Hubert a long time ago that he was poor. He was worth $8,000,000.

Free Road Maps

William B. Akin ran an advertising firm in Pittsburgh, Pennsylvania, and enjoyed riding in his 1912 Chalmers automobile.

But he often got lost on the country roads.

So he got an idea. "It would be a great help to motorists," he explained, "to have maps showing main roads of a county. Persons using such maps would be good prospects as gasoline customers. So why not combine the two features?"

That is just what Akin did. Akin called the general office of the Gulf Oil Corporation. Gulf president, G. R. Nutty, reached a quick decision.

Aikin's firm got the job of printing 10,000 maps of Allegheny County, Pennsylvania. A simple Gulf advertising notice was placed on the back. Then the maps were mailed out to the 9,500 registered automobile owners in the Pittsburgh area.

Free road maps made such a hit that now they are available at most gas stations.

Gummed Paper

Thomas Alva Edison is one of America's greatest inventors, famous for his development of the electric light bulb, the phonograph and many other major discoveries.

He was known to work tirelessly at a new idea until he came up with a workable answer. When he worked on the electric battery, for example, he completed more than 50,000 experiments before it worked properly.

Gummed paper was one of Edison's discoveries, however, that he didn't work hard on at all.

This is what happened. He had noticed that when he wanted to glue two pieces of paper together, he always got some glue on his hands; and it was unpleasant.

So he told an assistant to try leaving a strip of glue on some paper, then wet it slightly at a later time.

It worked. The moisture activated the adhesive, and the two pieces of paper stuck together.

Today postage stamps and envelopes use Edison's idea. All we need is a quick tongue to dampen the adhesive. No sticky fingers.

Helicopter

Helicopters were developed from windmills!

This isn't so surprising since windmills and helicopters both have large blades that turn in the air.

The first person to make windmill blades, which normally move in the wind, turn from an internal source of power, was the 15th Century artist, sculptor and inventor, Leonardo da Vinci.

Leonardo is famous for his painting of the Mona Lisa, but he also was way ahead of his time as an inventor. The helicopter is an example.

Leonardo figured that if he could make a "windmill in reverse," it would be useful. So he took a clock spring and fitted it inside a small toy-like object fitted with blades, or propellers. It worked. But what was he to do with it?

It was not until almost five hundred years later, after the development of the airplane, that engineers invented aircraft with rotary blades that could take off and land from a standing position.

American engineers, in 1919, had the technology. But it had been Leonardo, in the 1400's, who had the idea.

Home on Wheels

Arthur G. Sherman of Detroit loved camping, and it was summer again. The year was 1929.

He couldn't wait to get out to those wide open spaces, where there were no city noise and no crowds.

But one thing bothered him. He hated all the work of setting up camp every day, and taking it all down again.

He began to think of ways to make camping just as much fun, but less work.

He tried the one item then available which attempted to answer his need. It was a crude box on wheels, containing a folded tent, designed to be towed behind a car. Arthur G. Sherman didn't like it. It was almost as much trouble getting the tent out of the box as it was to get it out of his car.

Sherman got an idea. Why not go all the way and build an entire living space on wheels? Once back home after his summer vacation, Sherman went right to work, and for $500 he built a trailer with a room on it. It was nine feet long, and high enough for a grown man to stand in. It even had bunks and a coal-burning stove.

It was the world's first mobile home. Wherever it was towed, it caused a sensation.

Sherman went into business. He hired two cabinet makers, called his company the Covered Wagon Company, and built his trailers to sell for $400.

They sold like crazy. In only eight years, Sherman's stock in the company was worth almost two million dollars. *Fortune* magazine called it a "rather handsome reward."

Ice Cream Cone

For folks who lived anywhere near St. Louis, Missouri, the biggest event in the summer of 1904 was the Louisiana Purchase Exposition, which took place in that city.

No one knew beforehand, but that exposition was the occasion where ice cream cones were first made and sold.

The person who did it was named Charles Menches and he was a seller of ice cream. But he didn't plan to invent the ice cream cone.

This is how it happened.

Charles Menches sold his ice cream in dishes the way every other ice cream man did. That August when the Louisiana Purchase Exposition was at its height, was a real scorcher, however, and one day disaster struck Mr. Menches. There were so many hot and thirsty fair-goers wanting ice cream that he ran out of dishes.

And it wasn't even noon. He had more than half a day of business ahead of him and not a single dish to serve his ice cream on.

What did Menches do? He looked around him and thought fast. Nearby was a stand where his friend, Ernest Hamwi, who was from Syria, was selling a Middle Eastern treat called Zalabia. Zalabia consists of a crisp, wafer-like pastry sold with syrup.

"Give me Zalabia!" cried Menches. He rolled up the Zalabia, scooped his ice cream on top, and presto! ice cream cones were born.

Jeans

The Gold Rush of 1848 attracted many adventurers to California. One of them was a twenty-year-old named Levi Strauss.

Strauss had been a draper, or cloth seller, in New York, and he took a few bolts of cloth to sell on the journey west.

In this manner he earned his way, and by the time he reached California, Levi Strauss had sold everything except a roll of canvas.

No one wanted clothes made of canvas!

Or did they?

It turned out that "up in the diggin's," where the miners worked, pants wore out very quickly.

So Strauss made some pairs of canvas trousers to sell to miners.

More and more miners were coming to Strauss and asking him for a pair of those canvas trousers. It wasn't long before he forgot all about gold digging.

He called his pants "Levi's." They were popular with cowboys as well as miners.

Today, called Levis or blue jeans, they are popular with men, women, and children in many countries of the globe.

LEVI STRAUSS DRY

Jello

P. B. Wait was a cough syrup manufacturer in LeRoy, New York, whose business was not going very well.

So he decided to give up the cough syrup business and branch out into something new.

He picked the food industry. People eat all the time, he reasoned, while they take medicine only when they are sick.

For many years food manufacturers had experimented with gelatin, which is made from animal bones, but no one had been able to come up with a gelatin that was appealing. Gelatin looked bad, and it didn't taste very good, either.

So Mr. Wait went to work. His answer was to add fruit syrup to gelatin. He named his new product "Jello."

The new business had no competition, but, unfortunately, not enough people wanted to try Jello. So Wait ended up by selling out to a cereal manufacturer.

Later the cereal man tried to sell the Jello business, reportedly for only $35. And no one was interested in buying it!

About 1900 a number of cooking experts discovered Jello and decided it was just the thing for an elegant meal. That changed everything. Jello began to appear at banquets and fancy dinners. Today, Jello is known worldwide as a bonafide American food.

Jigsaw Puzzle

John Spilsbury was a geography teacher in London, England, who lived during the time of the American Revolution.

Spilsbury was always thinking of ways to be a better teacher. "How can I help my students learn faster and remember longer?" he wondered.

It was common in those days to use printed maps as a teaching aid. Spilsbury thought of something different.

He glued a regular map of England and Wales to a piece of wood. Then using a saw with a very small blade, he cut the map into pieces, following county lines.

He tried the puzzle with his students, who had fun fitting the pieces together and learned their geography in the process.

Soon many kinds of maps and pictures were being made, for amusement as well as educational purposes.

The puzzles were named after the "jigsaw" blade, the special saw which is still used for cutting jigsaw puzzle pieces.

Kodak

George Eastman bought his first camera in 1874, when he was twenty-four years old. It was to lead him to fame and fortune.

Photography at that time was not new. In fact, it had been over forty years earlier that the Frenchman Louis Daguerre, had invented the process. Professional photographers were already taking high quality pictures.

The trouble was that taking and developing photographs was very complex and very expensive. The cameras were large, and chemically treated glass plates were used as a recording medium.

George Eastman loved photography. He also believed that everyone should be able to enjoy it. He put his genius to work on the problem of making photography cheaper.

His first discovery was substituting coated film for glass plates. The film was light, flexible, and inexpensive. Then he developed a simple box camera that anyone could operate. It was made to hold enough film for one hundred exposures.

The first of Eastman's cameras, called Kodaks, went on sale in June, 1888. Loaded and ready for use, they only cost twenty-five dollars.

In the beginning, when a film was used up, the owner had to send it, camera and all, to the single developing plant in Rochester, New York, for printing. Soon, however, refinements enabled film to be printed locally.

The public loved that little Kodak camera, and Eastman's dream of popular photography came true.

THE NEW
KODAK
ROLL FILM
CAMERA
$ 25.00
You Press the button, we do the rest

Lawnmower

For centuries farmers cut grass and grains with a long, heavy blade called a scythe. It was hard work, but there was no alternative. Gardeners even used scythes to mow the fine-clipped, rolling lawns of aristocratic estates.

The idea for the first hand lawn mower came from the clothmaking industry and a man named Edwin Budding. Budding lived in a small town in England and worked in a cloth factory.

Much of the cloth produced in his factory turned out fuzzy and had to be trimmed smooth. A little machine had been developed for this purpose which consisted of revolving blades fixed between rollers.

Budding, who loved the outdoors, made the connection between trimming cloth and trimming lawns. He built a machine with longer blades and two wheels. He attached a long shaft so that persons could push the mower without bending down.

The year that Edwin Budding built his first lawn mower was 1831. At first, few people saw the point. The estate owners continued to hire gardeners to scythe their acres of lawn.

But the gardeners appreciated Budding's laborsaving idea. The popularity of lawn mowers grew, until every house with a yard about it was sure to have one.

Life Savers

In 1912, when candy maker Clarence Crane first marketed "Crane's Peppermint Life Savers," life preservers were just beginning to be used on ships—the round kind with a hole in the center for tossing to a passenger fallen overboard.

But that is not the whole story. Crane had been basically a chocolate maker. Chocolates were hard to sell in summer, however, and so he decided to try to make a mint that would boost his summertime sales.

At that time most of the mints available came from Europe and they were square in shape.

Crane was buying bottles of flavoring in a drug store one day when he noticed the druggist using a pill-making machine. It was operated by hand and made round, flat pills.

Crane had his idea. The pill-making machines worked fine for his mints, and he was even able to add the life preserver touch by punching a tiny hole in the middle.

Later Crane was to sell his rights to his new candy for under three thousand dollars. He may have regretted that decision, for "Life Savers" earned the new manufacturer many millions of dollars.

Lionel Trains

Joshua Lionel Cowen was an inventive guy. His electric flower pot had seemed like a good idea, but he gave it to a friend in order to work on the electric train (see FLASHLIGHT).

The fact is, Joshua had always been very interested in trains. When he was seven, he whittled a miniature locomotive from wood. It exploded, however, when he tried to fit it with a tiny steam engine.

Joshua had never forgotten his childhood experiment. Years later, he fitted a small motor under a model of a railroad flatcar, and the electric train was born.

He gave his middle name to the invention, and called his product, which at first consisted of the electric flatcar and thirty feet of track, the Lionel train. It cost six dollars.

Lionel trains proved to be very popular, and before long Joshua Cowen was manufacturing cattle cars, coal cars, passenger cars, train stations, and tunnels. It was 1910. Electric trains had become a big business.

Long Playing Record

Austrian-born Peter Goldmark was a scientist who loved classical music.

It was just after World War II, and Goldmark was visiting friends in Connecticut. His hosts said they had a record of Brahms' Second Piano Concerto—a favorite of Goldmark's—and put it on the phonograph.

Peter Goldmark was excited to hear the concerto because he had not heard it throughout the long, stormy years of the War.

Now in those days, all records were 78's. Each side lasted for about ten minutes.

Goldmark's enthusiasm waned with each stopping and starting. By the time they got to the third record change, he wished they had never started in the first place. How can one listen to Brahms and be interrupted every few minutes! he thought.

Later, he put his inventiveness to work in an effort to develop a way to listen to concertos and symphonies without interruption.

It took years of work. Eventually, however, Goldmark succeeded by producing a record with very narrow grooves. He also slowed down the revolutions per minute to thirty-three instead of seventy-eight.

The new "long playing" record totalled about forty-five minutes of playing time and revolutionized the record industry.

Malted Milk Powder

Malted milk, used today at soda fountains, was originally developed as a preservative for milk powder.

The man who discovered the method was William Horlick of Racine, Wisconsin.

Wisconsin is known as the "Dairy State," and in fact, produces so much milk that it has always been a problem what to do with the surplus.

Many Wisconsonites had tried to preserve milk. The most popular idea was to dry it until it became a powder. But milk powder suffered from one major draw-back. It tended to spoil during the drying process.

Horlick's idea was to add something to the milk that might prevent spoilage. He tried many substances. Finally, one worked: an extract of wheat and barley that had been treated with malt.

Horlick called his malted milk powder "Diatoid." People didn't like the name very much, but they liked his product.

Margarine

A century ago soldiers, sailors, and many others on the move often had to do without butter because it spoiled so easily.

Even salting, which was effectively used to preserve meat, did not help. Salted butter still turned rancid in a few days.

It was a French chemist who solved the problem. His name was Hippolyte Mege-Mouries. He used animal fat, or tallow, that could be mixed with milk and water, and the result was a butter-like product that stayed fresh for weeks and even months.

Mege-Mouries called the new foodstuff "margarine." Napoleon III's soldiers liked it, and so did many others.

The new substance became so popular that many nations who depended on selling butter for valuable income passed laws forbidding the manufacture of margarine.

Public demand proved too great, however, and today margarine is firmly established as a butter substitute.

Milk Cartons

Paper containers for milk are now so commonplace that it is hard to imagine that it took over twenty patient years of persistence to get the public to accept them. But it did, and the man who is responsible is named John Van Wormer.

Van Wormer owned a toy factory in Toledo, Ohio. He got the idea for paper milk cartons after dropping a milk bottle one morning. The bottle broke, the milk went everywhere, and it was an annoying way to start the day.

Van Wormer took out a patent in 1915 for his new product. He called it "Pure-Pak," because the container could be thrown away after one use.

Then the difficulties began. It took all of ten years for Van Wormer to perfect a machine capable of making the containers.

Even then the struggle was not over, for most Americans were very attached to milk in bottles. But time, and a steady increase in the cost of glass, worked in Van Wormer's favor.

Soon many other companies were making their own paper milk cartons, and by 1950, Van Wormer's company was producing them at the rate of 20,000,000 a day.

Miniature Golf

Toward the end of the 1920's golf was a popular American sport. Unfortunately, relatively few had access to a golf course and the opportunity to play.

John G. Carter of Chattanooga, Tennessee, had observed that many more people were interested in golf than played it.

Why not shrink the size of the course and make a miniature game? he thought.

He called it "Tom Thumb" golf, and this is why.

In the P. T. Barnum circus there was a midget known as General Tom Thumb. Tom Thumb married a midget, and their wedding achieved a great deal of publicity.

In fact, something about the story of Tom Thumb and his wife captured the popular imagination. For many communities throughout the country sponsored mock "Tom Thumb" weddings where small boys and girls, all dressed up, went through marriage ceremonies for fun.

Thus when John G. Carter invented miniature golf, "Tom Thumb" meant very tiny and lots of fun. It was a natural for Carter to call his new game Tom Thumb golf.

Interestingly, the name didn't stick, but the game did.

Monopoly

Clarence S. Darrow was a famous lawyer who lived at the turn of the century.

The inventor of the game Monopoly had the same name, but his middle initial was "B."

Clarence B. Darrow came from Germantown, Pennsylvania, and like many people, he never had enough money to suit his needs.

So he invented a game that let players imagine they were involved in big business. Monopoly players buy houses, hotels, railroads, and utility companies with play money. It is a buying and selling game, and big sums are involved.

Darrow made a number of sets by hand using linoleum and cardboard. His friends loved the game so much that Darrow began selling them commercially.

Later, he sold out to the Parker Brothers Company. Darrow had turned his idea, where players played at making a fortune, into a real fortune for himself.

Motor Bike

Gottlieb Daimler, a baker's son, got around his town, Wurttemberg, Germany, by bicycle. Perhaps because he was no longer young, he found that his legs often felt tired even after just half an hour of cycling.

There ought to be a better way to power a cycle, he thought. If only motors could be made to run more reliably, maybe one could be attached to a sturdy cycle.

Daimler had no training in engineering, but he correctly reasoned that a new fuel would have to be found to make his idea work. At that time most motors were powered by coal gas and needed a continuous flame. Daimler tried petrol, or gasoline, and ignited it with electricity.

Then Daimler reconstituted the bicycle frame. He used wooden wheels for strength, and fitted them to a heavy frame.

The year was 1884, and Daimler's invention was an instant success.

Neon

Georges Claude, who lived in Paris at the turn of the century, was the father of so many inventions that he is known as the "Edison of France."

About 1901 he found that a rare kind of gas, neon, could be obtained from the air. The problem was, in Claude's own words, "Once we had it, no one had any use for it."

Claude and his assistants experimented. The most interesting quality of neon, they found, was that it glowed when an electrical charge was passed through it.

This fact gave Claude an idea for an entirely new way of producing light.

He made neon tubes that could be used like ordinary light bulbs, but people did not take to the reddish light for household use.

Then by bending the tubes with heat, Claude found a way to form letters. He realized that neon tubes could be used as advertising signs.

The neon sign was first used in the United States in 1923 to advertise a play called "Little Old New York." The sign attracted more interest than the play, and the boom in neon had begun.

ORK
LITTLE OLD NEW YO
A GREAT MUSICAL!!
LITTLE OLD NEW YORK

Odd-Price Sale

Melville Stone was a self-made man who worked his way up from newspaper carrier to publisher of the Chicago *Daily News*.

When Stone first started his newspaper in 1875, the price was a penny. Circulation rose rapidly at first, then levelled off.

Then sales started lagging. When Stone investigated why fewer people were buying his paper, he discovered the problem had nothing to do with its quality.

Pennies were in short supply.

Stone decided he had to do something. First he traveled to the United States mint in Philadelphia and brought about the transfer of barrels of pennies to Chicago.

The problem then became how to get the pennies into circulation. So Stone persuaded Chicago merchants to sponsor "odd-price sales," during which they would sell their merchandise for a penny under the regular price.

The odd prices did the trick. People had pennies again, and Stone's paper flourished.

And that is why store items today cost "$8.99," or $12.99," instead of even dollar amounts.

Chicag
Daily
News

Paper Towel

Arthur Scott, head of a paper products company in Philadelphia, Pennsylvania, had big trouble.

An entire railroad carload of paper, unloaded at his plant, wasn't good for anything because the paper had been rolled too thick for toilet tissue, its intended purpose.

Was he going to send the whole load back?

Meanwhile, Scott heard about a certain teacher in the city school system who had developed a novel idea to help fight colds in school. She gave every runny nosed student a small piece of soft paper to use. That way the roller towel in the toilets wouldn't become contaminated with germs.

Scott decided he would try to sell the carload of paper. He made perforated rolls of "towels" thirteen inches wide and eighteen inches long.

That is how paper towels were born. It was to take many years, however, before they gained acceptance and replaced cloth towels for kitchen use.

Potato Chips

George Crum, an American Indian of the Huron tribe, held the post of chief cook at a resort hotel called Moon Lake House in Saratoga Springs, New York.

One summer a well-traveled guest made a fuss about the french fried potatoes. The pieces were much too thick, he complained.

Crum got angry. Just to spite the fusspot, he cut up some potatoes into wafer-thin slices, fried them up, and ordered the waiter to give them to the customer.

Instead of getting angry in turn, however, the guest thought the new fries were great.

Moon Lake House immediately began featuring the item on its menu as a new delicacy. Competing hotels soon began their own production of what came to be called "Saratoga Chips."

With the "Saratoga" long dropped, the specially thin fried potatoes developed into a whole new industry, for which sales in 1972 passed the one billion dollar mark.

Quonset Hut

When World War II began in Europe, American military leaders foresaw the day when the United States would be involved, and they began to prepare for the special problems of an overseas war.

Among other things they looked for a new type of building—one that could be put up quickly and be used in any kind of weather. The Great Lakes Steel Corporation of Detroit, Michigan, came up with the winning design. The company's engineers developed a prefabricated shelter that consisted of arching corrugated metal designed to be bolted to a steel foundation.

As a test, the first of the new shelters was built at a Navy base in Quonset Point, Rhode Island. The name of the base was the Quonset Naval Air Station.

Soon American fighting men in Iceland were sleeping in a building they called a "quonset hut," and before the end of the war, quonset huts were in use both in Europe and the Pacific.

Rubber Heels

Humphrey O'Sullivan was born in Ireland and came to the United States when he was twenty-one years old.

In Ireland he had been trained as a printer. So he found a job working in a printing plant.

Humphrey stood up at his work, often for twelve hours at a stretch, and the printing plant floor was made of stone.

No wonder he went home at night with aching feet!

To protect his feet from the cold, hard floor, Humphrey came to work one day with a little mat of rubber. The mat worked well, and he felt much better at the end of the day.

It worked so well, in fact, that his workmates kept borrowing his mat.

So Humphrey got the idea of making rubber foot-pads nailed on to the heels of his shoes. No one could take these, he reasoned.

In 1889, Humphrey O'Sullivan obtained a patent for rubber heels, and in not too long, he had his own business.

Today, hundreds of companies make rubber heels for shoes. But only those made by the O'Sullivan Company are advertised as "the world's #1 heel."

Safety Glass

Today safety glass, which will not splinter when exposed to shock, is everywhere—in windshields for cars, goggles for machinists, and windows and doors for many public buildings.

Essential as it is, safety glass was the result of a clumsy mistake.

Edouard Benedictus, a French scientist, was working in his laboratory. The year was 1903. By accident, he knocked a flask containing celluloid to the floor. Instead of flying over as dropped flasks usually did, this time the glass stayed in place. It was shattered, but it had not lost its shape.

Monsieur Benedictus knew he had discovered something, and he experimented further. Eventually he came up with a process consisting of sealing celluloid between two pieces of glass by heat.

You could hit it with a hammer and the glass stayed in place.

At first, no one was interested in safety glass. The age of the automobile changed the picture, however, and by 1914 cars began to be equipped with Benedictus' invention.

Safety Pin

Walter Hunt had no trouble thinking up new ideas.

First he invented a machine to spin flax. Then he invented a fire engine gong, a forest saw, a stove that burned hard coal.

His inventions worked, but he just did not have the knack for making money from them.

One day in 1849 Walter Hunt wanted to pay a fifteen-dollar debt to a friend. So he decided to invent something new.

From a piece of brass wire about eight inches long, coiled at the center and shielded at one end, he made the first safety pin.

He took out a patent on his invention, sold the rights to it for four hundred dollars, paid his friend back and had three hundred eighty-five dollars to spare.

Then he watched his latest brainstorm go on to become a million dollar money-earner—for someone else.

Self-Service Shopping

Clarence Saunders had lots of schemes for getting rich, but nothing seemed to work for him.

At the age of thirty-five, in 1916, Saunders was operating a small grocery store in Memphis, Tennessee.

That's when he got his inspiration.

Why shouldn't customers be allowed to walk through a store and select the items they wanted to buy? he wondered. A shop could serve more people that way, and carry a greater variety of products.

He promptly bought a stack of baskets for his customers to use as they circulated through his store serving themselves.

Rival grocers scoffed at Saunders' idea. First, they said, the customers would not like it. Second, Saunders would have a big stealing problem on his hands.

But the customers did like it—without stealing. Business proved so good Saunders opened a second store only five weeks later.

Saunders' concept was to sell a high volume of merchandise at a low profit. This meant cheaper prices. It was a brilliant approach.

Today the self-service supermarket dominates the grocery industry.

Self-Starter

Before Charles F. Kettering came along, automobiles had to be started by hand cranking the engine. It was hard work, dirty, and not a very pleasant or speedy way to begin a drive.

It was widely agreed that the solution to the problem was probably electricity. An electric motor capable of igniting a gasoline engine would be too heavy to put on a car, though.

Kettering's idea was to use a small electric motor designed to provide a big burst of power—but only for a very short time.

He completed sketches for his starter motor on July 23, 1910. Six months later he had built a prototype and successfully used it to start a Cadillac engine.

The problem was solved.

Sewing Machine

Elias Howe couldn't earn enough to support his family, and his wife took in sewing to make extra money.

Watching his wife sew one day, Howe got the idea for a machine that would sew.

It was much easier to think about than to build. It took Howe seven years, in fact, to make a machine that could sew in straight and even stitches.

In 1845, Howe demonstrated his machine, easily beating five women who were sewing by hand.

Nevertheless, no one would buy the new invention.

It was not until the Civil War produced a need for thousands of uniforms to be made quickly, that Howe's machine was adopted.

He went on to become a wealthy and famous man.

Shopping Bag

Walter H. Deubner ran a small grocery store in St. Paul, Minnesota, and he was looking for a way to give his business a boost.

By careful observation, he noticed that his customers' purchases were limited by what they could conveniently carry.

So he set about devising a way to help them buy more purchases at one time.

It took him four years to develop the right solution: a prefabricated package, inexpensive, easy to use—and strong enough to carry up to seventy-five pounds worth of groceries.

The package consisted of a paper bag with cord running through it for strength. Deubner named his new product after himself, calling it the "Deubner Shopping Bag," and sold it for five cents.

Deubner patented his product and within three years, by 1915, was selling over a million shopping bags a year.

Shopping Cart

Oklahoma City supermarket operator Sylvan N. Goldman watched, day after day, as customers struggled to carry all their purchases.

He decided to come up with a better way. Working in his small office one night in 1936, he placed together two small folding chairs.

"Inspiration hit me right between the eyes!" he later said. He figured out that by fastening the chairs together, putting wheels under them, and placing a basket on top of each seat, he had developed the shopping cart.

In Goldman's perfected model, steel replaced wood. His first metal cart is on permanent display at the Smithsonian Institution in Washington, D.C.

Tea Bag

Thomas Sullivan ran a small wholesale tea business in New York City at the turn of the century.

He had one major problem preventing him from expanding his trade. There was no convenient way for him to send samples of his many varieties of tea to prospective customers.

In those days, people made tea by brewing loose tea leaves in pots, then straining the tea into cups. If someone wanted just one cup of tea and didn't want to brew a whole pot, there were a number of devices he could use. One was called a "tea bell." Shaped somewhat like a bell, it held a small amount of tea and could be submerged in a single cup.

Sullivan decided to adapt the tea bell idea for sending out his samples. He made little bags of gauze, put different teas inside, then attached half a dozen of his "tea bags" to a piece of cardboard. This became his sample package, which he sent out to retail merchants.

To Sullivan's surprise, he began to get requests for shipments of the tea bags in quantity. The customers had seen the samples and wanted tea bags for themselves, it seemed.

Sullivan complied, and tea bags were soon available for household use.

Teddy Bear

Theodore "Teddy" Roosevelt, president of the United States from 1901-1909, was a famous outdoorsman and game hunter.

Once on a hunting trip in Mississippi a bear cub trotted out from a clump of bushes directly in front of the President. His companions shouted at him to shoot. But he couldn't do it. Something about the little cub had touched him.

Later a cartoonist for the *Washington Star,* Clifford K. Berryman, drew a cartoon showing the President sparing the life of the "Teddy Bear."

Toymakers immediately took up the idea and began to manufacture Teddy Bears. At first, the stuffed bears were made to look like the President. Soon, however, Teddy Bear became the standard name for any stuffed bear, whether it looked like Teddy Roosevelt or not.

Telephone Directory

Alexander Graham Bell's telephone was an instant success in the popular imagination. But few actually paid money to have a telephone installed in their homes or business.

By 1878, for example, New Haven, Connecticut was one of the biggest telephone centers—and there were only fifty subscribers.

By today's standards, fifty is a small number. It was enough, however, to inspire someone working for the New Haven Telephone Company to suggest compiling a list of the numbers of every subscriber.

The suggestion was considered so unimportant at the time, that no one recorded whose idea it was originally.

The telephone company did compile a directory, though.

A century later telephone directories have become a vital element in world communication. Every locality has its own directory, painstakingly revised each year, and the phone book for a big city, like New York or Los Angeles, may have as many as four volumes.

New Haven
Directory
of
Telephone
Subscribers

Traffic Light

Even during the horse and buggy days, traffic in big cities was often heavy. Police officers had to be stationed full time directing traffic at busy intersections.

And with the coming of automobiles, the situation got even worse.

Police Officer William Potts of Detroit, Michigan, decided to do something about the problem. What he had in mind was figuring out a way to adapt railroad signals for street use.

The railroads were already utilizing automatic controls. But railroad traffic traveled along parallel lines. Street traffic traveled at right angles.

Potts used red, amber, and green railroad lights and about thirty-seven dollar's worth of wire and electrical controls to make the world's first traffic light.

It was installed in 1920 on the corner of Woodward and Michigan avenues in Detroit.

Within a year, Detroit had installed a total of fifteen of the new automatic lights. Today, there are over 125,000 throughout the United States.

648

Traveler's Checks

James C. Fargo, president of the American Express Company, was well off and well known. It's not surprising that he felt insulted when he couldn't get checks cashed during a trip to Europe in 1890.

But the European bankers were steadfast. Fargo was not known to them. So they would not cash his checks.

Was an American always going to have a cash problem when traveling in Europe?

An employee of the American Express Company, Marcellus F. Berry, set out to find a solution. He wrote later:

"There's one thing every person does in a distinctive way. That is writing his signature. Therefore the foolproof device for taking money to strange places must carry the signature of the bearer. It must declare that it will be cashed only when a second, and matching, signature is added before witnesses."

On July 7, 1891, Berry was granted four copyrights for what he called "the traveler's cheque," and William Fargo, James Fargo's son, got the first one. He had no difficulty when he wanted fifty dollars a few weeks later in Leipzig, Germany.

In 1891, American Express sold $9,120.00 worth of traveler's checks, and the amount has risen every year.

Typewriter

When Christopher L. Sholes of Milwaukee, Wisconsin, had worked as a newspaper editor, he became interested in trying to perfect a "writing machine." Many had tried, but none had succeeded.

While serving as collector of customs under President Lincoln, he began experimenting with a machine that could write. He did develop something, but it was clumsy and did not work very well.

It was observing a pianist play the piano that gave Sholes his inspiration. He arranged a set of keys such that when one was pressed, a lever caused a metal letter to strike a roller.

Thus the typewriter was born, although it took Sholes seven years to perfect the first model, which he completed in 1868.

Mark Twain was the first author to write a book with a typewriter. He called his machine "the curiosity-breeding little joker."

Umbrella

The umbrella is almost as old as written history. We know about it from records over 3,000 years old from Asia.

Europeans did not become acquainted with the umbrella until the Twelfth Century, when it was brought to Venice.

Umbrellas were used as sunshades as well as protection from wet. At first they were considered no more than a novelty, and their spread to Northern Europe was slow.

It was due to an English merchant, Jonas Hanway, that the umbrella became popular. He was a rich trader in Asian goods, and he had come to find his umbrella very useful. Starting in 1750, he decided to take a daily stroll with his umbrella in order to demonstrate it.

He walked through fashionable London, on sunny and rainy days, with his umbrella always on show. He ignored the snickers of those who thought he was crazy.

Jonas Hanway went strolling with his umbrella for thirty years, and as time went by more and more of the residents of rainy London saw the usefulness of the device.

Today, of course, the umbrella is a standard item throughout the world, and especially in London, where we always associate the Englishman with his umbrella!

Vacuum Cleaner

Hubert Booth was an English bridge builder at the turn of the century.

One day he went to see a new kind of cleaning device in use at a London railroad station. The device used compressed air to blow the dirt towards one side of the floor.

The machine was an improvement over a broom, but cleaning men still had to sweep up all the dirt by hand.

Booth decided a sucking device would be more efficient than the blown air device. His engineer friends belittled his suggestion; it had already been tried, they said, and the dirt had clogged up the machinery.

Booth was not put off. He tried a simple experiment. He put a handkerchief over his mouth, lay down on a dirty floor and sucked in air.

Sure enough, the handkerchief caught the dirt.

He constructed a machine that sucked air, equipped it with a filter, and the vacuum cleaner was born.

Vulcanized Rubber

Today we ride in cars equipped with rubber tires because Charles Goodyear had a lucky accident.

Before 1839, rubber could not be used to make tires because it changed in elasticity according to the temperature. Goodyear was interested in "curing" rubber so that it would have a wider usefulness.

He had already tried mixing it with all sorts of things, such as ink, black pepper, sulfur, even cream cheese. But the rubber was still brittle in winter and sticky in summer.

One day Goodyear went to get some groceries and accidentally dropped a small piece of crude rubber on a hot stove used to heat the store.

Instead of melting, as he might have expected, the rubber hardened and at the same time remained pliable.

Goodyear's rubber specimen had contained sulfur, and it was heat in combination with sulfur that did the trick.

From the name of the Roman god of fire, Vulcan, Goodyear called his treated product "vulcanized rubber," and rubber tires became a reality.

Windshield Wiper

When Mary Anderson of Birmingham, Alabama, visited New York City in 1902, there was one thing that she didn't like—the electric trolleys.

It wasn't that they were noisy, or too crowded. Mary thought they were unsafe. Numerous times, while riding in the trolley in bad weather, she observed that the driver could hardly see out the windshield.

In those days, trolley drivers couldn't do much to keep rain and snow off their windshields. Rubbing the outer surface with moist tobacco or an onion put a slight film of oil on the glass which helped a little—but not much.

Safely back home in Alabama, Mary thought she could remedy the problem. Her solution was to attach a handle to a rubber blade. A driver could turn the handle back and forth by hand and make the blade wipe off rain and snow.

As is often the case with new inventions, many people who heard about Mary Anderson's "window cleaning device" thought it was silly. But drivers thought it was great.

Soon windshield wipers were standard equipment on many sorts of vehicles.

Xerox

When Chester F. Carlson, who worked for an electronics firm in New York City, needed something copied, he had to take it to a typist or a photographer.

Sometimes it could take days to get a simple document copied.

Carlson decided to make a machine that would produce copies instantly. He saw it in his mind: a person would feed a document into the machine, push a button, and seconds later receive a copy. The year was 1935.

He went to the library and did research. He discovered that experiments had been carried out in Hungary in which light and static electricity were utilized to duplicate images on paper.

Using this principle, Carlson began to build his copying machine. Three years later he copied his first document successfully. He called the process Xerox (meaning "dry writing") and tried to sell it to a major corporation.

Not a single company was interested, however. Finally, with the help of a non-profit institute, Carlson was able to develop the first commercial Xerox machine.

Xerox machines became available to businesses in 1950. Carlson's years of work paid off: the Xerox corporation became one of the biggest successes in American business history.

Yale Lock

In the 1850's most locks were crude and easy to pick open. Locking up a house or a box of valuables so that it was burglar proof was often an impossible task.

It was a man named Linus Yale, Jr., who changed the situation by developing a brand new principle in lock construction. His discovery was so important that his name became known in many countries, and his lock completely replaced the old-fashioned models.

What did the Yale lock consist of? Inside was a cylinder, into which was fitted a key. When the cylinder was turned, the key touched a set of tumblers controlled by springs. A specific pattern of grooves on the key tripped the tumblers in such a way as to trigger the bolt of the lock. If the key did not have the right grooves, the tumblers would not move the bolt.

What assured Linus Yale of success was that the new Yale lock was almost impossible to pick. It was burglar proof.

Few people know that Yale got his idea for the tumbler system from the ancient Egyptians, who developed locks on the same principle over two thousand years ago!

Zipper

Whitcomb L. Judson loved machines and experimented with many different kinds of gadgets.

He invented a number of labor-saving items, including the zipper. It came about because of a friend's stiff back.

The problem was that his friend could not do up his shoes. Judson came up with a slide fastener that could be opened or closed with one hand. This was an absolutely new idea, and in a few weeks Judson had a working model.

On August 29, 1893, he patented his new "hookless fastener." But it was not until twenty years later, when the B. F. Goodrich Company decided to market galoshes with hookless fasteners, that the product became popular.

These new galoshes could be fastened with a single zip of the hand, and soon hookless fasteners came to be called "zippers."